The Mozzle of the Milky Way

Rajarshi Sarma

Copyright © 2020 Rajarshi Sarma
All rights reserved.

ISBN-13: 9781650802077

Price: $ 20 USD

DEDICATION

To Lord Vishn

Preface

I would like to convey my special thanks to my Parents for encouraging me in every steps of my Life. Since I was kid I used look up into the night sky and tried to figure out why the stars are dazzling. Even I always thought there is an earth out there. Perhaps that is why I am inspired to write something about astronomy. This is a small piece of work. I hope I am able to summarize some contents inside this book. Milky way and the whole Universe is that thing which can never be summed up inside a book. Still I had to add so much in it but I had to finish it up within 15 days. I am planning to add more researchable things in my next astronomy book. While I was writing this book I have added some pictures which are taken by NASA. So special thanks to NASA.
I hope readers will love this work. Any fruitful suggestion is welcome.

Author

Rajarshi Sarma

Introduction

The sky is infinite. You have no limit up to which you can explore this amazing infinite distance. If you look up into the sky that means you are travelling in the time dimension . You see the great Andromeda galaxy in the night sky and you are looking at 2.5 millions of years back. If you see the Altair star of our home galaxy in the night sky that will take you 16 years back in past. Even you are looking at the sun every day and you are looking at past tense by 8 minutes. Guess why? That's the time trap in which we all are trapped in. There's no escape from it. The space is a part of very difficult mathematics which mankind will never fully understand. In the present observable universe in each and every second numerous stars are born and die, galaxies are merged or/and separated and so many activities are taking place that we are never aware of. Since the beginning of the big bang the universe is expanding. The universe consists of matter and antimatter. Just like white has no existence in absence of black ; matter has no existence in the absence of antimatter. Are these antimatter forming another anti universe? Not a fiction but thinkable. If it is then it must have been 14 billion years back in time. But if you travel to that hypothetical anti universe you will be exploded as human body consists of matters. On upon exposure to anti universe matter and antimatter will get merged and release tremendous energy and disappears without a trace. Really spooky?

Anyway we shall discuss these later on but first let me introduce you to our home the solar system.

The Solar System

The solar system is controlled by its host star the sun. it is G type star and it is the only star of this stellar system. There are 8 planets known till now and Pluto is considered as dwarf planet because there are numerous Pluto like objects in the

Kuiper belt. Beside planets and star sun there is an asteroid belt between mars and Jupiter. Of course the comets are the guests of solar system which are believed to be the native of oort cloud.

The solar system has certain boundary recently the voyager 1 and 2 mission has discovered so many interesting facts about the edge of the solar system. According to NASA's recent update pressures is so much higher in boundary of the solar system. The pressure is due the force of plasma, magnetic field and particles that collide with each other and exerts pressure. The sun shoots the plasma into the space which travels in space in much faster speed. This is called solar wind. The Earth has good magnetic field that is why it can protect its atmosphere from this strong Solar wind. The solar wind ends at the termination shock and after that the Heliosheath continue to Heliopause and after that interstellar space begins.

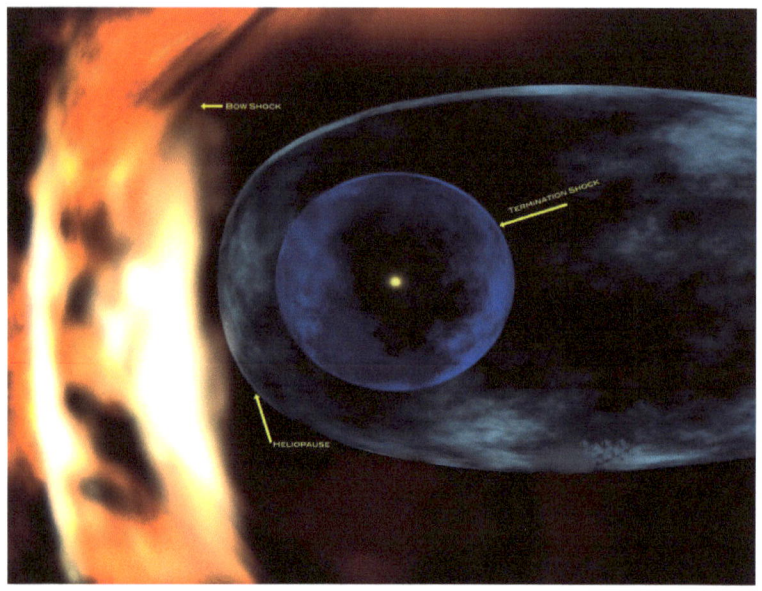

Now let's begin with the planets of the solar system.

Credit: NASA showing the boundary of the solar system

<u>Mercury</u>

*Image: Recent picture of mercury
Credit: NASA*

Mercury is the first planet of the solar system. You may think it is the hottest planet of the solar system but it is not

As observed on 26

December 2019 at 1:42 pm the following data has been summarized-

Mercury

- Type- Planet
- Magnitude- -0.73
- Absolute magnitude- -0.60
- R/A DEC. (J2000) : 17 H 39 M 0.7.89S
- R/A DEC. (On date) : 17 H 40 M 22.73 s
- Mean sidereal time: 20 h 06 m 14.2 s
- App. Sidereal time: 20 h 06 m 28 s
- Paratactic angle: 37.55°
- IAU Constellation: oph
- Distance from sun: 0.465 AU
- Orbital velocity: 39.052 kmS^{-1}
- Equatorial rotation: -0.003 KmS^{-1}
- Apparent diameter: +0°00'04.76"
- Equatorial diameter: 4879.4 Km
- Sidereal period: 87.97 Days
- Mean solar day: 4222h 27 m 52.55 s

- Phase angle: $18°45'00.9"$
- Albedo: 0.060

This image shows the craters present on the mercury surface. It is the closest one from the sun. It was named after the Roman God Mercury. Interesting fact is that it takes only 87 days to orbit around the Sun. It is tidally locked to the Sun in a 3:2 resonance that's why it takes so much longer time to rotate around its own axis. The 3:2 resonance means it rotates on its axis 3 times for every 2 revolution around the Sun. Its orbital eccentricity is the largest known in the solar system. If you go to mercury your birthday will be in 87 earth days but one day will be 116 Earth days.

Atmosphere and Surface:

Mercurian atmosphere is really thin as it is too close to the Sun. The Sun also looks four times in size from the surface of the mercury that we look from the Earth. The atmosphere consists of atomic oxygen, sodium, magnesium, atomic hydrogen, potassium, calcium, helium, traces of iron, argon, carbon di oxide, water vapor, xenon, krypton and neon. Atmospheric pressure is less than 1 Nano Pascal.

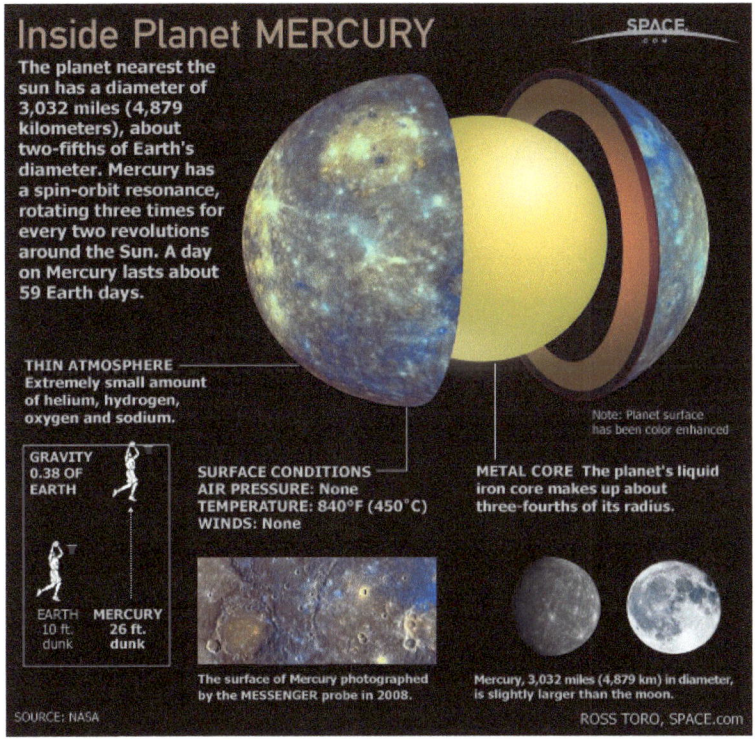

Mercury is the rocky planet. It has hills, rocks, craters on its surface but all are lifeless. 70% of the body made up of metallic and 25% silicate and rest other trace s of elements. It is the smallest planet of the solar system and has no natural

This image shows the physical property of the planet

Source and credit: NASA

satellite. Even though there are so many hypothesis are there about its formation but no one knows the real fact about it. The space probe MESSENGER took so many interesting photographs of the planet. The largest known crater is Caloris basin. There are so many terrains can be found on the surface of the planet. Some other known craters are Tolstoy basin, abedin crater. These were formed due to either asteroid or comet impact with the planet. The impacts on this planet are frequent as sun attracts every space rocks towards it so Mercury suffers a lot.

Surface Temperature:

One must think that being closest to sun it must be the hottest one. But practically no. It is tidally locked to sun so one surface stays for 116 Earth days in front of the sun and other surface suffers night for 116 Earth days. So the surface temperature ranges from 100K to 700K.

In the polar regions its temperature always below 150 k. Even though NASA has confirmed large ice cap at the north pole of the planet but due to extreme variation of temperature it is impossible to support life on the planet.

Venus

Venus is the living hell of the solar system. It is thought that once it supported life but due to closer proximity to sun it loses its water and become lifeless due tremendous GH affect

*Source: NASA
Venus as captured by NASA probe*

Venus:

- Type: Planet
- Magnitude: 3.94
- Absolute magnitude: -5.18
- R/A DEC (J2000): 20H 41 M 32.99S
- R/A DEC (ON DATE): 20H41M22.83 s
- H/A DEC: 23 H 53 M 41 S
- Gal. long./lat.(J2000): +307°11'32.5"/-1°51'43"
- Mean sidereal time: 20 h 38 m 22.9 s
- Apparent sidereal time: 20h 38m 40s
- IAU Constellation: Cap
- Distance from sun: 0.727 AU
- Orbital velocity: 34.8 KmS^{-1}
- Equatorial diameter: 12103.6 Km
- Sidereal period: 224 earth days
- Mean solar day: 2802 H
- Phase angle: +47°55'33.4"
- Albedo: 0.77
- luminosity: 83.5%

This above image is the only image ever taken on the surface of the Venus. It was taken by a soviet lander Venera 13. The atmospheric pressure of Venus is way too high that a human will be crushed by this pressure and its temperature can melt the hardest metal found on Earth. Venera was the first lander and it worked for 3 hours on the surface and then it got disabled. This color photo shows Venus surface got rocky terrains and it has sulphuric acid rich atmosphere. No stars can be seen in sky even the Sun can never be seen due to its thick atmosphere. Venus is a hell. Heat is always trapped inside due to huge greenhouse effect and temperature is around 500^0 C.

Observation can be done if you look at evening sky right after Sun set and the second brightest heavenly object in sky after Moon is the Venus. Generally it is seen in the west.

It takes 224 Earth days to orbit around the Sun. but it's one rotation on its axis takes even longer i.e. 243 Earth days. Interestingly Sun rises in the West in Venusian surface.

This sister planet of Earth doesn't contain any natural moon. It has very dense sulphuric acid cloud in its atmosphere. Solar wind often sweeps away the hydrogen from the atmosphere. It surface pressure is 9 MPa. Atmosphere contain 97% CO_2 2.5% N_2 and rest consists of Water vapor, SO_2, Ne, HCl, HF, He.

80% of the Venusian surface is made up of smooth, volcanic plains, ridges. Alpha region, beta region are some highland regions. There often lightning and thunder can be seen. Even sulphuric acid rain is prominent. However recent studies has shown some evidence of similarity of its atmosphere with Earth's current atmosphere billions years ago. But the runaway greenhouse affect causes severe irreversible changes which may had wept out any civilization from the planet. Be careful human we are doing same thing on Earth too.

Pentagram Of Venus:
It is a path that Venus make while orbiting around sun as observed from the Earth.

There are not much space programs has been launched to explore this planet but still in 1974 mariner 10 approached Venus and took several images. In 1975 venera 9 and 10 were launched but in 1982 venera 13 took the only color photo from its surface. Recently messenger (2006) is the latest project. However Japan launched Akatsuki is orbiting around the planet since 2015.

Interesting fact is Venus has still habitable earth like condition in air about 50 miles above from the surface. Guess what a town in the sky is possible but still not much oxygen.

Mars

Mars is the most explored planet by human. It is however very interesting planet. Mars was discovered and named after the Roman God of war. Surprisingly early observatory reported an unknown symbol looks like 'W' on its surface. Even great scientist Nicola tesla wanted to establish connection with Martian civilization if any by sending signals. One evening he received signal from outer space and thought it was from the mars but later on it was revealed that this was radio wave tested by Marconi not by outer space civilization.

Anyway this can be our second home but it is impossible to colonize human there via our present technology. First let us see the geography and other facts about this planet.

Mars
- Type: planet
- Magnitude: 1.59
- Absolute magnitude: 1.52
- Mean opposition magnitude: -2.01
- R/A DEC (J2000): 15H 34 M32 S
- R/A DEC (On date): 15h 35m42s
- Mean sidereal time: 4h 08m 13s
- Apparent sidereal time: 4 h 08 m 23s
- IAU Constellation: Lib
- Paralactic angle: -44.9^0
- Distance from sun: 1.59 AU
- Orbital velocity: 23 km/s

- Equatorial rotational velocity: 0.241 km/s
- Apparent diameter: 0°00'04"
- Equatorial diameter: 6792 Km
- Sidereal period: 686.6 earth day
- Sidereal day: 24h 37m
- Synodic period: 779.95 earth days
- Phase angle: 23°36'34"
- Illuminated: 95%
- Albedo: 0.15

Mars has drawn so many attention from human. Many thinks even I think that Martian civilization perhaps influenced the human civilization on Earth. Maybe we are decedent of Martian civilization who left mars during their last moment but it looks like a fiction story yet it has been strongly recommended to study about it. Anyway let's talk about the missions by human first

First mars explorer program was launched in 1960, 10 October. It was **1M No. 1** spacecraft which was operated by **OKB-1** by Soviet Union. It was a flyby mission which was failed to orbit. After 4 days of launching **1M No1** another spacecraft **1M no.2** was launched and it was a failed mission too. Then in 1962 there were two missions launched by **soviet union-2MV-4** and **Mars-1** both were terribly failed. Then after soviet union's back to back failure missions American space agency NASA launched its first flyby mission to mars on 5 November 1964. It was **mariner -3** mission which was A FAILED MISSION TOO. But human never give up. NASA launched another **Mariner 4** on 28 November of the same year and it was the fly by mission too. The rockets involved were **Atlas LV3 Agenda-D**. This was the first successful mission by human race towards any planet. It reached mars on 15 July 1965. Soviet union also launched another spacecraft **Zond2** in the same year of **mariner 4** launched. But they lost communication with the spacecraft and it was lost. In 1969 NASA and soviet union launched couple of mars missions each. All were flyby missions but both of Soviet union's missions were failed.

NASA however was successful Mariner 6 & 7 missions. In 1971 NASA had a failed mission **mariner 8**. But soviet union had its first successful mars mission in 19 may, 1971 **Mars-2** spacecraft which was an orbiter carried by rocket **proton-k/d**. After that few mars missions were launched by both NASA and Soviet Union but in 1975 the significant **Viking-1, 2** landers and **Viking 1,2** orbiter. **Viking 1** was the first lander on mars. However soviet union had most failure missions and after formation of Russia the last mission **Mars 96** was launched and it was failed too. In 1996 Sojourner was the first rover sent to mars by NASA. NASA launched the very popular mars missions in 2003 which was Spirit and Opportunity. Both were Rovers. They landed on Mars on January 4, 2004 and 25 January 2004 respectively. They were really amazing rovers which clicked so many images that human had never expected to see on Mars. They worked like hero and alone they were working years on Martian surface withstanding the harsh Martian weather. It was operated by solar energy. The team of spirit and opportunity had to face so many problems while they were operating them. Usually signals sometimes in a midway delay between 4 minutes to 24 minutes. So during some crucial events like if the rover got stuck into the soil of Mars team found real difficulties. The rover was not designed to tilt its solar panels hence during winter it was needed to place it in a such way that its solar panel is facing towards the sun in order to supply energy to the battery. They sent so many images to earth. Even though they're machines they worked even better than a human. In 2004 European space agency launched a gravity assist names **Rosetta**. Then NASA's **phoenix, curiosity, Mars reconnaissance orbiter, Maven, Insight** were the landers that launched between 2005 to 2018. The other countries who has taken part in mars mission are Japan, India, China. However 100% success rate by ISRO India. ISRO launched Mangalyan in 2013 which was an orbiter. It is still operational.

Mars 2020 is a highly anticipated mission by NASA in July, 2020 which is supposed to fly robotic helicopter in the sky of Mars. Apart from it 2020 MBRSC of UAE has proposed too launched Hope Mars Mission which will be an orbiter. Exomars by Russia and European space agency is another proposed mars mission. CNSA, China will launch MGRS orbiter and rover in 2020. In 2022 Japan will launch Mars terahertz microsatellite. ISRO India is set to launch Mangalyan 2 which is rover, lander and orbiter. Martian moon explorer will be another mission to study its moons by Japan in 2024.

Possibility of sending human to mars pretty challenging. To reach mars around 8 months are required. For an astronaut to travel in a small cockpit of a spaceship for 8 months is really difficult. The astronaut maybe mentally unstable for a continues travel. Moreover the threat from cosmic rays from sun is really harmful for human. Prolonged weightlessness may cause eyesight impairment and blood flow including blood clotting malfunction in human body. Potential failure of life support is the main issue which will be fatal.

Now let's talk about –the features of this amazing planet.

It is the fourth planet of the solar system. It is smaller than the earth. The Iron oxide is abundant on the surface which imparts the red color to the planet hence it is called red planet. It has two moons phobos and deimos. These two are small moons and they orbit around the planet. Phobos and deimos has low albedo value due which it is thought that they can be asteroids which trapped to Mars magnetic field many millions years ago and became its moon. Phobos orbits around the planet near the synchronous orbit. Due the tidal force gradually phobos is lowering towards Mars and it is predicted that after few millions of years it will crash into the Mars or it will break apart and its debris will form a ring around the planet. Deimos orbits around the planet via a larger orbit. Phobos rises in the west and sets in the east and it rises again after 10 h 50 minutes. One Martian day is 24 hours which is similar to Earth but 1 Martian year is about two earth years. The Temperature is extreme on Martian surface. In some areas like two poles and northern hemisphere the temperature at night can fall up to -143^0 C to -20^0 C. during day times the temperature maybe around 0-20^0 C. But in some places it maybe 35-50^0 C during daytime. The greenhouse effect is very poor in Martian atmosphere. Although it is composed of CO_2 BY 95%. The atmosphere has thin layer of ozone as well. Other gases found in Martian atmosphere are oxygen, neon, water vapor, nitrogen, hydrogen deuterium oxide, krypton and xenon.

Geology of Mars:

It is a rocky planet. Tall hills, hillocks, craters are common in Mars. During NASA's mars mission there are some interesting facts that were revealed. Olympus Mons is the tallest known mountain of the whole solar system. It is taller than mount Everest. Valles Marineris is largest known canyon in the solar system. The Northern hemisphere has covered largely by Borealis basin. the mars is near the asteroid belt so it was prominent target by asteroids in past. Aitken Basin is largest impact crater found on Mars. It is maybe created by an impact really giant asteroid or comet. Which must be four times in the size of our Moon's poles.

Spirit rover found silica rich dust on

Martian surface. Phoenix lander sent few data about the soil of Mars which indicated it is alkaline slightly and contains Mg, Na, K, Cl. Martian soil P^H is found to 7.7 and it is toxic because it contains perchlorate salt. Dust devil is the biggest threat to human exploration on Mars. NASA lost its Rover due huge dust storm which acquire speed up to 150 Km/h. This dust storm can spread though whole planet and it can break human body into pieces if someone's trapped inside the storm. Current climate of Mars cannot hold water on its surface in liquid form as its low atmospheric pressure. However NASA's Mars reconnaissance had sent some data which shows that a very large amount water is in the form of ice is present on the two polar caps of the planet. If it is melted it is sufficient to cover the whole planet with water up to a depth of 30 ft. As observed by the rovers it is clearly understood that liquid water was present on its surface once. Martian surface has huge linear swathes of scoured ground. These outflow channels is as large as the amazon or Brahmaputra of India. Such one example is Ma'adim Vallis. It is 20 Kilometer wide and with 1km average depth. Perhaps it was big river of Martian surface millions years ago. Recently it is found that continues evolution of methane gas has been detected. This maybe sign of life on Mars. But if it is then it will be microbes or there is some microbes from earth which somehow managed to reach mars via landers and contaminated the Mars. But landers are well sterilized. So it is interesting from where these Methane are coming from. Secondly if Mars had water in past it had suitable atmosphere like earth. Which means it supported life. Perhaps the civilization was so civilized that it destroyed themselves by wars or they got wept out by asteroid impact or intergalactic war destroyed them. These may look like fiction but truth is in between. Some unofficial evidence has claimed that they found cannon type things on Mars. Even some metallic things were claimed to be found.

Image shows sunset in Martian Sky
Credit: NASA

Image shows Earth seen from night Martian sky

If Mars had life then it was lost either due to nuclear war, intergalactic war, global warming, asteroid impact or due loss of Martian magnetic field due which drastic solar wind weeps away the molecular oxygen and carbon di oxide from its atmosphere making atmosphere less dense. Whatever it is, still I must say terraforming Mars through nuclear bombing on its both poles as proposed by Elon Musk of SpceX may not be a good idea. Because it may have some negative impact if there is some underground creatures. Secondly we don't know what affect earth may get from it because we cannot underestimate this mysterious universe. Thirdly it will cost so much money so instead of spending on terraformation of Mars better spend it for the poor children who sleep every night on the road without eating a single bread.

Curiosity and Rover on Mars.
Credit: NASA

Jupiter

Jupiter is the 5th planet of the solar system. It is largest gas planet.

Jupiter the largest planet with its giant red spot

Jupiter:
- Magnitude: 1.84
- Absolute magnitude: -9.40
- Mean sidereal time: 4h 22m 59 s
- Apparent sidereal time: 4 h 23m 10s
- Paralactic angle: 77.2^0
- IAU Constellation: Sgr
- Distance from sun: 5.222 AU
- Equatorial rotation: 12.57 km/s
- Orbital velocity: 13 km/s
- Apparent diameter: $0^0 00' 31"$
- Equatorial diameter : 142984 Km
- Sidereal period: 4331 days
- Sidereal day: 9h 55m
- Mean solar day: 9 h 55 m
- Synodic period : 398 days
- Phase angle: $0^0 05$
- Illuminated: 100
- Albedo: 0.510

Jupiter can be seen from earth with naked eyes. It is really bright star. After moon and Venus Jupiter is the brightest.

Jupiter 79 total moons as reported in 2018.

Atmosphere of Jupiter:

Jupiter's atmosphere contains 90% H_2. Its atmospheric pressure is 20-200 KPa which is enough to crush any human if lands on its surface. Other gases are Helium, methane, ammonia, Hydrogen deuterium, ethane, water. However these are present in traces of amount except helium 10%. Jupiter may have rocky core but it lacks solid surface. As seen in the images Jupiter has repetitive bands on its atmosphere. The great red spot is the most significant that is observed since it was discovered. This great red spot is greater in size of the earth. Jupiter is covered with ammonium hydrogen sulphide cloud. As mentioned above clouds are arranged in bands.

Jupiter is the home of strong storms. Wind speed may exceed 360Km/Hour. Jupiter has the most powerful magnetic field among the planet. If we could see the magnetic field it would be seen bigger than our moon in the night sky. Because of strong magnetic field it acts as shield against the incoming comet and space rocks towards Earth. We should be thankful towards Jupiter. Surrounding of Jupiter there is a faint ring.

There were several missions towards Jupiter. Out of that voyager flyby, pioneer flyby and Galileo orbiter are significant. Recent orbiter sent to Jupiter is Juno an orbiter which has entered in its orbit in 2016.

From Juno mission it is found that the cloud layer is only 50 Km deep. There are two decks. One is thick and other is thin. Juno has also detected lightning in the atmosphere.

The Moons:

Although Jupiter has 79 Moons we study only few of them. Out of them Io, Europa, Ganymede, Calisto are to be studied. These were first seen by Galileo hence called Galilean Moons.

Jupiter and its main four moons

Io:

Over 1.8 Earth days, Io rotates once on its axis and completes one orbit of Jupiter, causing the same side of Io to always face Jupiter.

It is slightly larger than Earth's moon.

The moon Io is the most volcanically active world in the solar system. Io even has lakes of molten silicate lava on its surface.

Io's very thin atmosphere is primarily sulfur dioxide, which on Earth is sometimes used to preserve dried food.

Io has no known moons of its own, but it is possible for moons to have moons.

Io has no known rings, but it does create a gaseous torus of material along its orbit around Jupiter.

Spacecraft have studied Io on flybys

(Voyager and Cassini) or orbiting Jupiter (Galileo). Most recently, New Horizons observed Io en route to Pluto.

Io almost certainly could not support life as we know it. But that's not to say it couldn't harbor some form of life as we don't know it.

Io's volcanoes are at times so powerful that they are seen with large telescopes on Earth.

Image shows Io
Credit: NASA

Europa:

This moon of Jupiter may support life. It is thought that it contains ocean subsurface below its icy layer.

Smaller than Io but it is believed it can support life. It

Europa 6th largest moon of solar system Credit: NASA

has smoothest surface. The heat of tidal flexing may cause the ocean of ice in liquid water state. Even though if there life exists it will be unicellular life. Complex life is not possible in Europa. Its surface temperature is average -160^0C which is not suitable for human. Rossby waves causes significant influence in Europa.

Ganymede and Calisto are other two important moons. Ganymede is larger than Pluto and Mercury and it is the most massive and largest Moon. It has own magnetic field. It also contains water ice and silicate rocks. Voyager 1 and 2 provided so many data regarding this two moons which are still analyzed. **Jupiter Icy Moon Explorer** will launched in 2022 by European Space Agency. It will examine more about these moons.

Saturn

Image of Saturn captured by Cassini Credit: NASA

Saturn is the 6th planet of the Solar system. It is very beautiful because it has an amazing ring around it. Which is supposed to disappear after few millions of years.

Saturn is gas giant it has the highest moon of the solar system. Total 83 moons it has. The moons are regular or irregular. Some are of few kilometers of diameter some are bigger.

Saturn as observed on 31 December 2019

- Magnitude: 0.54
- Absolute magnitude: -8.88
- Mean sidereal time: 4h 29m 59 s
- Apparent sidereal time: 4 h 29m 10s
- Parallactic angle: 74.52^0
- IAU Constellation: Sgr
- Distance from sun: 10.034 AU
- Equatorial rotation: 9.871 km/s
- Orbital velocity: 9.17 km/s
- Apparent diameter: 0000'31"
- Equatorial diameter : 120536 Km
- Sidereal period: 10760 Earth days
- Sidereal day: 10h 39m

- Mean solar day: 10 h 39 m

- Illuminated: 100%

- Albedo: 0.500

Saturn has density lower than Earth but it 95% more massive than Earth. Surface gravity is 10.4 m/s². This is almost similar to 9.8 m/s² of Earth. Saturn has average temperature of -150°C on its

Saturn as observed by Cassini. Saturn cloud has been seen Credit: NASA

surface. However its surface pressure is huge 140kPa. Unlike Jupiter also consists of Hydrogen mainly. Helium, methane, ammonia, ethane are other gases consists the atmosphere of Saturn. Saturn core is surrounded metallic hydrogen, liquid hydrogen. Saturn has weaker magnetic field than earth. Its magnetic moment is around 500 times of earth's because of its large size. Saturn hosts tremendously speedy wind. But the wind speed is less than Neptune's. Wind speed may exceed 1800 Km/h. Imagine if this kind of wind break out in earth. A cronian is of 10 hours only. Great white spot is a unique phenomenon of cloud and wind which occurs in every 30 earth years.

Saturn is famous for its ring system. The ring system is nothing but some icy particle, dust and rocks. It is assumed there are millions of ampere current is travelling across the rings. The ring extends from 6630-120700 Km. Jupiter and other gas giants also has ring but Saturn's ring is prominent. Saturn's ring maybe formed due to remnants of a destroyed moons. Or, due remnant of nebular material the rings was formed. Probability is in between.

Escape velocity of Saturn is around 36 Km/s along the equator zone.

Saturn has 11700^0C temperature in its core. It has no correct explanation but it is thought that due to raining out of Helium droplet in deep Saturn core. Rainfalls of diamond can be suggested in Saturn.

Now let us talk about the natural moons of Saturn. Amongst the 83 moons we need to explore Dione, titan, mimas, Tethys, enceladus, Rhea.

Titan is the largest Moon. It is the most suspected moon to support life. Titan has dense atmosphere. Surface pressure is

146 kPa. Atmosphere consists of Nitrogen, methane, hydrogen.

Titan is rocky moon contains ice, ammonia rich liquid water. Its dense opaque atmosphere may cause problem in studying interior of titan. Titan contains ocean of liquid methane and it supports methane cycle. There is methane rain can be seen. Surface temperature is -179^{0}C.

Image: Titan the largest moon of Saturn.
Credit: NASA

Rhea is the second largest moon of Saturn. It was discovered by G. Cassini in 1672.

Rhea as shown by Cassini probe. Credit: NASA

Surface temperature 53 K covered by ice and cannot support life.
The other moon enceladus may support microbial life but it is not sure. It has ocean with salt rich particle. Methanogenesis occurs in this moon also

a potential sign to support life that lives by Methanogenesis.

There were several explorations done by NASA. Pioneer 2 was first fly by mission. Then Voyager 1 and 2 were the other fly by missions. Cassini-Huygens spacecraft is the most significant mission. It entered in Saturn's orbit in 2004. It took 8 years to reach the planet. Cassini captured so many images of the planet and its moons. Cassini has tracked powerful lightning in Saturn. Huygens space probe crash

Image shows predicted look from the atmosphere of titan

landed in titan's surface and Cassini crashed into Saturn after clicking a mesmerizing Last picture of Earth from Saturn. We salute those robotic probes.

Uranus & Neptune

Neptune is the 8th planet of the solar system. Third most massive planet is an ice giant.

Image showing Neptune Credit: NASA

It is a near twin of Uranus.

Neptune

- Magnitude: 7.92
- Absolute magnitude: -6.87
- R/A Dec (J2000): 23H10M
- R/A Dec (On date): 23 h 11 m
- Mean sidereal time: 3h 22m 59 s
- Apparent sidereal time: 3h 23m 10s
- Paralactic angle: 57^0
- IAU Constellation: aqu
- Distance from sun: 29.8 AU
- Equatorial rotation: 5 km/s
- Orbital velocity: 5.4 km/s
- Equatorial diameter : 49528 Km
- Sidereal period: 60189 earth days
- Sidereal day: 16h 55m
- Mean solar day: 16 h 55 m
- Synodic period : 367 days
- Phase angle: $+1^005$
- Illuminated: 100
- Albedo: 0.62

Neptune is fourth largest planet. The

escape velocity on its surface is 23 Km/s. so once we land on its surface it will hard for us to get back to our home. Its solar day is about 16 hours while it completes one orbital rotation in 60189 earth days. Its north pole declination is 42^0. Surface temperature is average -218^0 C which completely unfavorable for living world. Composition is almost same as Jupiter and Saturn. It contains maximum percentage of hydrogen then helium and relatively lower volume of methane, hydrogen deuterium, ethane. Discovery of Neptune was based on mathematical calculation. A scientist Bouvard had notice unusual changes in the orbit of the Uranus. He then believed that perhaps some another planet which causes this gravitational perturbation. Then few scientists in 1846 observed Neptune by telescope. Currently 14 moons are known however Triton was the first discovered one. Due to its very long distance from earth it is hard to predict things. Voyager 2 visited Neptune in 1989. Neptune has dynamic storm. Which the most speedy storm of the solar system. Its speed around 2200 km/s. the south pole suffers day for 42 earth years and night for 42 earth years. Seasonal changes occurs in Neptune.

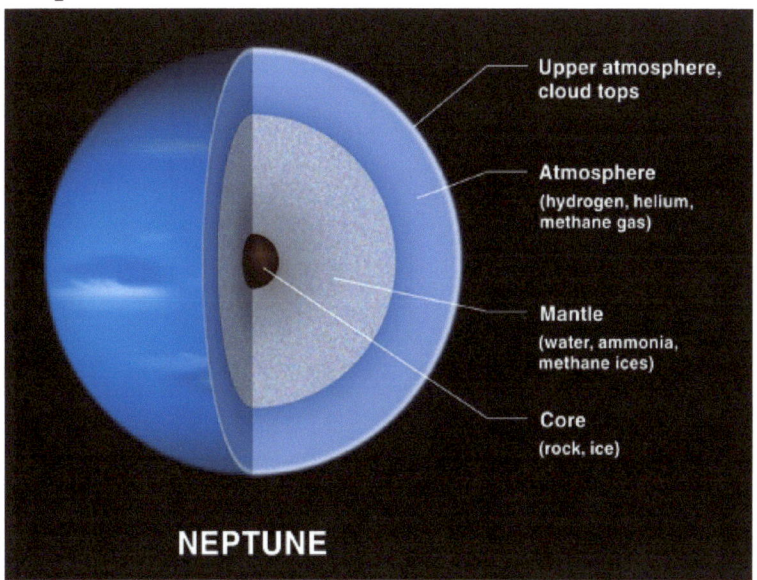

Core of Neptune Schematic image

Moons of Neptune

Out of 14 moons triton has retrograde orbit. Triton is believe to be planet trapped by Neptunian magnetic field.

Image showing moons of Neptune Credit : NASA/Voyager mission

The other satellites Nereid, proteus, naiad, thalassa, despina, Galatea. In 1981 Larissa another moon was discovered. There is a planetary ring system possible in Neptune. Rings are believed to be consists of icy particle coated with carbon material giving reddish hue.

In case of exploration only voyager 1 and 2 are the flyby that passes Neptune. New horizon technique is highly anticipated to explore Neptune in future.

Uranus

Uranus is the 7th planet. It is almost similar to Neptune and is ice giant. It cannot be seen from earth with naked eye. Its characteristic composition almost same to that of Neptune. Having stormy weather it cannot support life. According to voyager 1 and 2 images it has no bands and completely featureless. It has 27 known moons. Its rotational speed, sidereal time, solar day almost same as that of Neptune.

Umbriel, Miranda, titania, Ariel, Oberon are five significant moons of Uranus.

Uranus as clicked by Voyager
Credit: NASA

Uranian satellite system is not that massive. The total mass of five moons of Uranus is less than Neptune's moon triton. Titania is the largest moon of Uranus. But it has diameter of nearly 800Km. The moons are composed of ice and rocks. Ice is not of water. That ice is of include ammonia and carbon di oxide. The moons have impact crater, canyons on its surface. As Ariel is newly formed it has less crater. Umbriel has more crater as it oldest. Miranda has large canyon of 20 km wide.

Ring system of Uranus is not that prominent as Saturn or Neptune. Uranus ring is composed of dark materials. They may vary in diameter. Scientist William Herschel first described the Uranian ring. When voyager 2 passed Uranus it took clear picture of Uranian ring and this concluded that total number of rings in Uranus that is 11. In 2006 certain new observatory suggested that the outermost ring is blue and the inner is red and grey.

The explorations towards Uranus are not frequently done. Only the voyager 2 probes passed Uranus in 1986. Voyager 2 on 24 January 1986 made its closest approach to Uranus and studied its weather and chemical composition of its atmosphere. Many outstanding images were sent to earth including its rings.

Uranus is very far from earth. It will take around 30 years to reach Uranus however future Space mission like Uranus Pathfinder and some other has been proposed.

Kuiper Belt & Edge of the solar system

Kuiper belt also known as trans Neptunian belt is in the border of the solar system consist of the remnant particles which formed the solar system and its constituents. The constituents maybe icy volatile objects. Through New horizon technology so many studies has been done. After Gerard Kuiper its name had been kept. It orbits the sun and it is orbited by Pluto. From Neptunian orbit it is about 30 AU distant away. From Sun it is 50 AU away. Three dwarf planets Pluto, Haumea and Makemake are found in the Kuiper belt. It is similar to that of Asteroid belt but it is 200 times massive than asteroid belt. Previously it was believed that Pluto is a planet but now it is confirmed that Pluto is actually a Kuiper belt object (KBO). Pluto is the largest and most massive KBO. Although I have mentioned that Kuiper belt is the border of the solar system but actually is much more smaller than oort cloud in diameter. Oort cloud is the outer most layer of solar system which maybe a home of numerous comets. Oort cloud is 100 AU in diameter and it must be elliptical. The objects found in between Kuiper belt and oort cloud are termed as trans Neptunian object. The comets that we see in the sky are actually a traveller to the inner solar system from the oort cloud.

The resonance of the main region of the Kuiper belt is 2:3. But at the distance between 40-50 AU resonance is roughly 1:1.65. Due to gravity of Neptune there are some destabilization due to which Kuiper belt is scattered in certain regions. In the asteroid belt there is similar gap which is known as Kirkwood gap.

Heliosphere

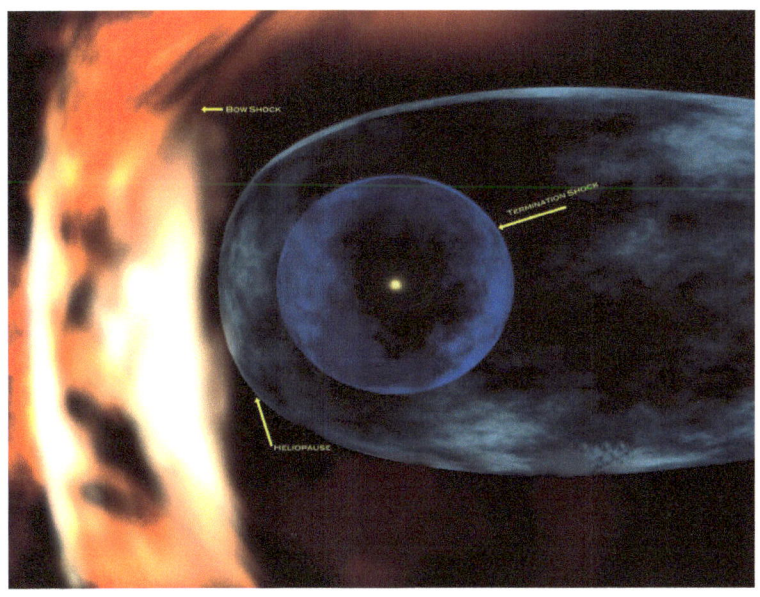

Schematic diagram of Heliosphere

Heliosphere in other words a spherical boundary of the solar system. Solar wind is a stream of plasma particle shot by Sun at an average speed 600 Km/s. this solar wind travels up to termination shock where it encounters with interstellar wind and this result in decrease in its speed. However the solar wind emanating from the sun creates a bubble and this is the Heliosphere. As the sun orbits around the giant blackhole of the milky way galaxy this Heliosphere also runs with the sun.

Heliosheath

Heliosheath is the outer region of the Heliosphere. In this region the Solar wind piles up and become slowest and hottest. Pressing against the Interstellar wind.

Heliopause

This is the region at junction between the solar and interstellar wind. This causes a certain amount of stable pressure which helps in backflow of the solar wind towards the tail of the Heliosphere.

Bow Shock

Bow shock is formed because of the plowing of Heliosphere through the interstellar space.

Heliosphere is most important for us as it shields from the harmful interstellar radiation. The only human made things ever reached the Heliosphere are Voyager 1 and 2.

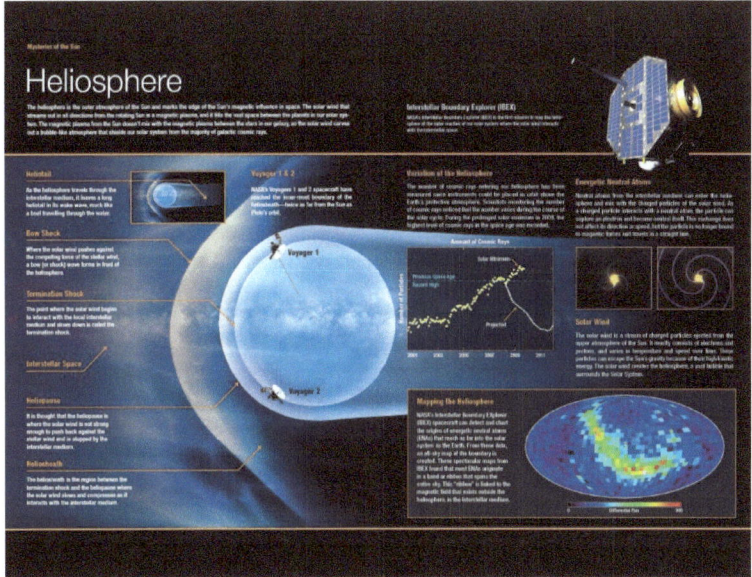

Alpha Centauri

Alpha Centauri is the nearest star system of us. It is roughly only 4.2 light years away from us. As it is the nearest star system of Solar system we have more chances to find a habitable exoplanet in this star system. Scientists have found an exoplanet in that stellar system. Alpha Centauri is a triple star system. The three stars are α-Centauri-A, α-Centauri-B and α-Centauri-C. α-Centauri-A is also known as Rigil Kentaurus, α-Centauri-B is known as Toliman, α-Centauri-C as Proxima Centauri. Before peeping into its interior we must look at its history. 1603, Johann Bayer suggested this star system. It has its traditional name Rigil Kentaurus. The second name is Toliman and third name is Bingula. In 1915 α-Centauri-C was discovered by Robert. International Astronomical Union (IAU) in 2016 approved the names Rigil Kentaurus to α-

Centauri-A, Toliman to α-Centauri-B, Proxima Centauri to α-Centauri-C.

What is this triple star system? This three stars are bright in the night sky except one i.e. Proxima Centauri. Proxima Centauri is a red dwarf star. The other two stars Rigil and Toliman are Sun like stars. They can be categorized as K type star.

Morgan-Kiernan Classification of Stars:

Type- O

Type- B

Type- A

Type- F

Type- G

Type- K

Type- M

From the hottest O-type to the coolest M-type. Now the temperature of each spectral class is then further divided into hot to cold. From A0 to A9 is one type sub classification. Similarly from B0 to B9 and so on for each type.

A0 will be the hottest type A star. A9 will be the coldest type-A star. Sun is a G2V type star. Its luminosity class is V (Roman letter). Further discussion about the stars is included in later chapter.

Rigil Kentaurus:

- Type- Double star
- Magnitude- 0.10
- Absolute magnitude- 4.45
- Color index (B-V)- 0.60
- R/A Dec (J2000)- 14 H 39 M 52S
- R/A Dec (On date)- 14 h 39 m 51s
- Mean sidereal time- 7 h 14 m 48 s
- Apparent sidereal time- 7 h 14 m 49 s
- IAU constellation- Cen

- Distance from earth- 4.39 Light Years
- Spectral type- G2V
- Parallax- 0.742"
- Position angle (2014)- 279^0
- Separation (2014)- 4.33"
- Proper motion by axis- -4607.0 486.2 (Mass/ Year)
- Position angle of the promotion- 276^0

Proxima Centauri (α-Cen, C-V645, Cen-HIP-70890-SA0147855, HD-9300-WDS, J14396-6050AC) :

- Type- eruptive variable star, Double star (UV)
- Magnitude- 11.0
- Absolute magnitude- 15.44
- Color index (B-V)- 1.80
- R/A Dec (J2000)- 14 h 29 m 37s
- R/A Dec (On date)- 14 h 31 m 38 s

- Eclipsed obliquity (On date)- +23°26'10"
- Mean sidereal time- 8 h 10 m 26 s
- Apparent sidereal time- 8 h 10 m 27 s
- IAU constellation- Cen
- Distance- 4.22 light years
- Spectral type- M6Ve
- Parallax- 0.77233"
- Position angle (2000)- 212°

α-Centauri- A has almost the same mass and luminosity of the Sun. α-Centauri-B is less luminous and α-Centauri-C is a red dwarf star. Age of these three stars are thought to be relatively same. The estimation is known as chromospheric activity. It is found that our Sun is little younger than these three stars.

α-Centauri-A is the primary of this stellar system. It is yellowish in color stellar classification is G2V type. α-Centauri-B is secondary star in this triple star system. It is K1 type star. Energy is released in X-ray band. α-Centauri-B is more magnetically active than Sun. According to Kepler's law there must be at least one planet in each stellar system. This stellar system also contains exoplanet. Amongst the exoplanets there must be a planet in the habitable zone of the parent star. So if we look at that particular spot we must have more chances to find alien life. Amongst the billions of stars in the Milky Way there are more than billions of planets. Where roughly more than millions of planets in the habitable zone. So it must be possible that we have millions of alien life present in our galaxy and some of them must be too advanced or moderate. The interstellar distance is very much large. It is measured in terms of light time. If a distance is traversed by light in 1 minute it is termed as 1 light time. If lights takes 1 hour then it is 1 light hour and if light takes 1 years of time to travel a certain distance it is termed as 1 light year. Light is the fastest known thing in the universe. From Einstein's relativity theory it is believed that there is no object in the known universe is faster than light. So if someone want to go to the α-Centauri system from Earth at present technology it will not be possible for him to reach the destination during lifetime. Light travels at 300000 Km/s. There is 31,536,000 seconds in 1 year. So in 4.22 years there is 133,081,920 seconds. Light will travel 3.9924×10^{13} kilometers in 4.22 years. If you go to this stellar system with the fastest spacecraft till now made by NASA that travels with speed 73 Km/s then you will need 17,342 Years to reach that stellar system. So it is unreal to think about these missions with our present dwarf technology. Anyway now we need to understand how these exoplanets are discovered. The only planet that is confirmed in this α-Centauri system was discovered by the radial velocity method. In 2016 it was confirmed. We have four major techniques to trace an exoplanet in our galaxy.

1. Astrometry
2. Doppler spectroscopy or radial velocity method
3. Gravitational micro lensing
4. Transit method

Astrometry- it is a method by which reflex motion of the star in the place of the sky is measured.

Radial velocity- it uses the Doppler shift of stars as it orbit the center of mass.

Gravitational Micro lensing- when a star passes of a background of star gravitational micro lensing occurs.

Transitional method- when a planet crosses in front of the host star there is variation of the light spectrum which indicates the presence of a planet around the star.

The planets are revolving either around the whole star system or one the three stars. The only discovered planet of α-Centaury system is revolving around the Proxima Centauri the red dwarf star. Still it is not confirmed if it is tidally locked to the host star or not. This planet is known as Proxima Centauri-B. the distance between this planet to its star is roughly around 0.05 AU. As Proxima Centauri is a red dwarf so it radiates less energy toward this planet. But the stellar wind is really drastic towards this planet so that it may erode it's atmosphere. Another probability is that it is may be tidally locked to Proxima Centauri. So if its eccentricity is 0 then it's one surface will be always facing the Star and other surface will be always facing the dark side. So in this extreme condition even though this planet is in the habitable zone there will be no life. There is no proper picture of this planet except few artist's impressions. There is no proposed exploration for this stellar system till date.

Milky Way Galaxy

Milky Way galaxy is our home in this universe. It is appeared as hazy band of light in the night sky. From our planet Galileo Galilee first discovered that the band of light is actually formed from the stars. It is a barred spiral galaxy with 200 Kly diameters. Thickness of the stellar disc is about 2 Kly and it contains around ±400 billion stars. Its spiral pattern rotation period is 220-300 millions years and bar pattern rotation period is 150 million of years. The escape velocity at Sun's position will be 600 Km/s that's why to escape the galaxy is impossible for human so there is no real image of the whole Milky way galaxy in practical. We have the image of the great Andromeda and other galaxies but in case of our home galaxy we only have its galactic center's image as seen from our earth. We actually see the supermassive blackhole of more than 4000 million solar mass. This supermassive blackhole is really massive one and it can eat any stars at any time. Recently scientists have observed a Sun like star got eaten by this massive blackhole. It is believe that many billion years ago the milky way galaxy had died and it lost its star forming gases. Later on it regained its capacity to form the stars and now we have our present shape of this galaxy. But there are few stars which may not be originally formed from Milky way galaxy. If that is true then perhaps our galaxy had merged with some foreign galaxy in past. Similar mega event will happen to our galaxy again when the great Andromeda galaxy will have an head on collision with our galaxy. Now let's talk about what really Andromeda galaxy is?

Andromeda is a spiral galaxy that is about 2.5 million light years away from Earth. It is measured that the redshift is -0.01±0.000013 and the minus sign shows that it is blue shifted. The Andromeda galaxy is approaching towards the milky way at speed 300 km/s and it is predicted that it will collide with our home galaxy in 4.5 billion years. The predicted beautiful sky is shown in image. There will be a beautiful sequence of

Image showing The Great Andromeda galaxy

events.

The first image shows the current position of Andromeda galaxy and Milky Way. Then in 1.5 billion years the sky will look like the seconds image. The third image of sky is of about after 4.5 billion years. So during this time the both galaxies will start colliding each other. This collisions will last for many million to billion years. But then to watch these spectacular events the human civilization will no more exist. Even our Earth will be eaten up by the Sun because in 4-5 billion years Sun will expand its core and it will passes its diameter beyond the orbit of Saturn and become a red giant.

After this mega collision this two galaxies will be merged in one galaxy and there will be no free existence of milky way and Andromeda. The massive black holes of the two galaxies will be merged or one will be expelled out and the most massive one will be the new blackhole of this newly formed galaxy.

Andromeda has two satellite galaxies. It is believed that these satellite galaxies will be merged too. Even the NASA or any scientists haven't confirmed the existence of alien civilization in anywhere till date but I strongly recommend that there is surely at least 1 advanced civilization in each galaxies of the universe. Andromeda is bigger than Milky way so obviously there is at least one advanced civilization which perhaps peeping into the night sky watching our milky way approaching towards them. Two hot neutron stars have been discovered in the Andromeda galaxy till date. Its two satellite galaxies M32 And M110 may also have civilization which may or may not be advanced.

Satellite galaxies of Milky Way:

Large Magellanic Cloud: Large Magellanic Cloud is 163000 light years away from our home galaxy milky way. It orbits around the Milky Way and it is a dwarf spiral galaxy. Generally it is observed from southern hemisphere of Earth. In night sky you can see it as a faint and smaller dust like appearance.

Small Magellanic Cloud: it dwarf irregular

galaxy and can be observed from Australia and other regions of southern hemisphere. Its distance maybe around 190-200 kilo light years. There may be a mini Magellanic cloud exist.

Image showing Magellanic Cloud

Rajarshi Sarma

Author

Rajarshi Sarma is an Indian Author from Guwahati, Assam. This is second published work and first book in astronomy. He studied BSc. (Hons) Electronics in University Of Delhi. Currently he is student in medical science (MBBS) in Tezpur Medical College and Hospital, Assam. His other book : The Cursed Island: The crooked man and the mermaid kingdom.

Readers can provide their valuable feedback and connect the author via instagram, Facebook.

Instagram handling: @theauthorfictio

www.ingramcontent.com/pod-product-compliance
Lightning Source LLC
Chambersburg PA
CBHW040319220526
45473CB00009B/2489